The Changing Arctic Landscape

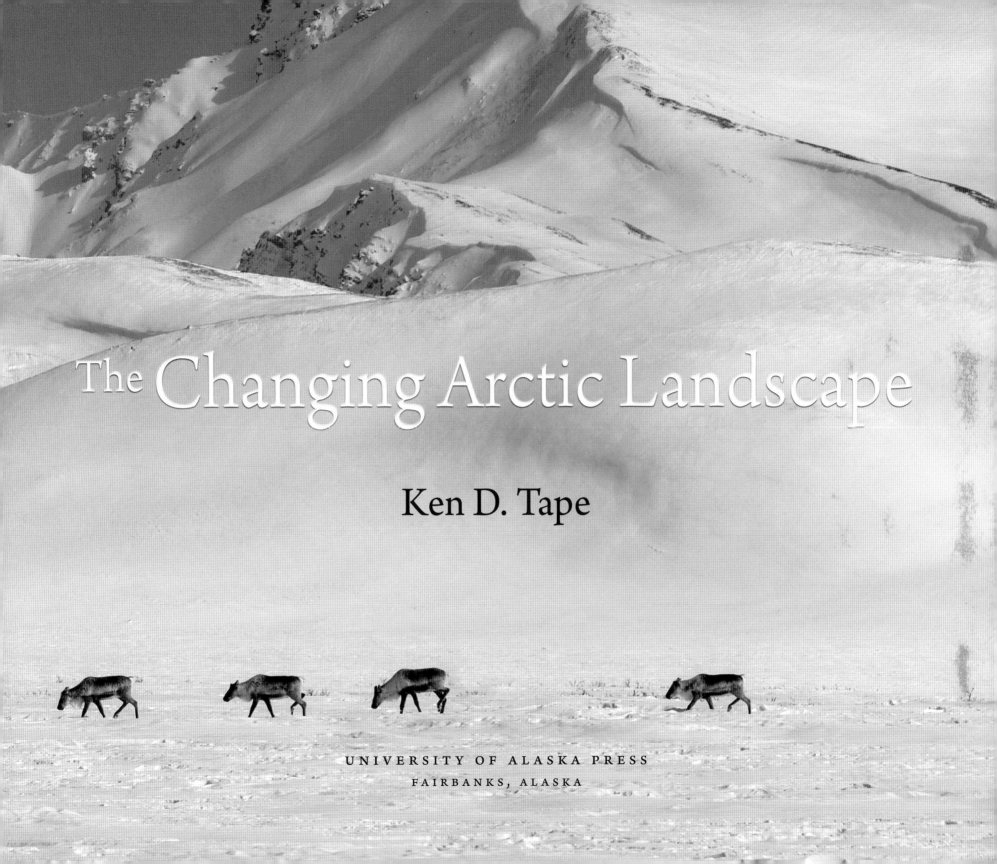

The Changing Arctic Landscape

Ken D. Tape

UNIVERSITY OF ALASKA PRESS

FAIRBANKS, ALASKA

P.O. Box 756240
Fairbanks, AK 99775-6240
www.uaf.edu/uapress

Printed in China

This publication was printed on paper that meets the minimum
requirements for ANSI/NISO x39.48–1992 (Permanence of Paper).

Library of Congress Cataloging-in-Publication Data

Tape, Ken.
  The changing arctic landscape / by Ken D. Tape.
      p. cm.
  Includes bibliographical references and index.
  ISBN 978-1-60223-080-4 (cloth : alk. paper)
  1. Alaska--Climate. 2. Climatic changes--Environmental aspects--Alaska--Pictorial
works. 3. Environmental conditions--Alaska--Pictorial works. 4. Tape, Ken. I. Title.
  QC984.A4T37 2010
  551.69798--dc22
                        2009035478

Book design by Dixon J. Jones, Rasmuson Library Graphics

Contents

Sunrise at 3 A.M. at Jago Lake, North Slope.

Preface

SOME SCIENTISTS WORRY THAT the changes now underway in arctic landscapes portend larger and perhaps sinister changes in the near future, both in the Arctic and elsewhere. The purpose of this book is not so much to speculate on the implications of the current changes but rather to document them. As evidence for these changes, I offer fifteen pairs of photos of sites scattered across northern Alaska. Each pair consists of an old photo—several decades old, at least—together with a recent photo of the same site taken from a perspective as close as possible to that of the old photo. By carefully comparing the old and new photos in each pair, you can see for yourself whether change is occurring. In some pairs the change is blatant, in others it is subtle, and in still others there is no change at all. Because the photographed sites are widely scattered, the collection of photo pairs as a whole is probably representative of northern Alaska and thus will permit some generalizations about landscape change in the region.

My approach is partly autobiographical. I wanted to convey some of my own appreciation for arctic Alaska and its history. I also wanted to convey some of the excitement of searching for old photos in archives and elsewhere that would be suitable for repeating. I wanted to share the inspiration that I got from the takers of those old photos, especially since those photographers were often pioneers in the study of Alaska landscapes. I also wanted to say something about the challenges and excitement of repeating the old photos. Finding and accessing the old sites was usually an adventure in itself;

successfully locating an old site was cause first for exhilaration, then for humility and awe at what the original photographer must have gone through.

Because the book has an autobiographical slant to it, the treatments both of the history and of the science are uneven in places, and neither treatment is meant to be exhaustive. Especially in the case of scientific contributors, I have naturally been drawn to those who were more closely associated with the areas of the repeat photos.

I was lucky to be able to meet three scientists who had done early work in the localities of some of the repeat photos, and in fact these men were the sources of some of the old photos that I later repeated. Their reminiscences in several interviews breathed life into the old photos. Therefore, I included in the book some of the biographical information gleaned from those interviews.

This work was supported in part by the National Science Foundation Office of Polar Programs, the Environment and Natural Resource Institute at the University of Alaska Anchorage, the United States Geological Survey (USGS) Alaska Science Center, and the University of Alaska Office of Academic Affairs. Additional support was provided by Carleton College, the National Park Service, the Arctic Research Consortium of the United States, and the Center for Global Change and Arctic System Research. The USGS photo archive in Denver, Colorado, generously granted me access to old photography. I would like to thank Tom Osterkamp, Vladimir Romanovsky, Grant Spearman, Matthew Sturm, and Matt Nolan for their many helpful discussions and suggestions. Thanks to Dixon Jones for his expert book design. Thanks especially to George Gryc, Ed Sable, and Art Lachenbruch for sharing their experiences and for enlightening me about the Arctic.

Figure 1   A sky full of ice crystals, late April, arctic Alaska.

Introduction

A VISITOR STANDING ALONE on Alaska's arctic tundra is apt to be struck by the seeming timelessness and constancy of the place. And indeed there is scientific evidence to suggest that, until recently, the landscape had not changed much over the past several thousand years. In the absence of fire to occasionally reset the vegetation, landscape change had been largely confined to seasonal changes and fluctuations from one year to the next. A deep snow year, for example, might increase ground temperatures, but the following year a shallow snow cover would return the ground temperatures to their previous state.

My acquaintance with the Arctic was forged in a hurry, after gliding over the Brooks Range mountains in a small ski-plane and landing softly on the broad, treeless tundra known as the North Slope of Alaska. After a few moments of dumping gear into the snow, the plane was gone, leaving only two wide ski tracks on the snow-blanketed landscape. I stood there in the cold stillness. The other three expedition members were veterans of this country, but they, too, paused for a moment, awestruck. It was –37°C (–35°F), and the clear air was filled with ice crystals that were forming colorful arcs across the sky (**Figure 1**). I had an overwhelming sense that time—particularly on the scale of years, centuries, and millennia—was no longer relevant.

The high temperature on that journey came on the last day of sampling, just south of the village of Barrow, when temperatures climbed to a balmy –31.6°C (–25°F). Being a total novice in the Arctic, and unprepared except in attitude, I suffered on that trip, spending most of it sleepless and dodging or succumbing to frostbite. It was remarkable, though, how quickly my selective memory forgot the suffering and displaced it with the beauty and timeless simplicity of the snow-covered Arctic.

In subsequent years I returned to the Arctic, drawn in part by the apparent constancy of the place. This was a landscape that had been relatively unchanged for six thousand years and would likely remain so for the next six thousand years. Or would it?

Since the late nineteenth century, the climate has been rebounding from a cool period known as the Little Ice Age. It is thought that this was a Milankovitch (orbitally) induced cool period from about 1500 to 1880 AD. Since then, temperature records show that the global Arctic has experienced several especially warm periods, including the latest warming, which began in the early 1970s and continues today (**Figure 2**). This warming has been more pronounced in some regions, like arctic Alaska, where temperature records show about a 3°C (5°F) increase over the last half century (**Figure 3**). What impact does this warming have on the arctic sea and landscape? Does the planet act as a buffer to these changes, absorbing and enduring climate modifications without change at the Earth's surface? Or, does the surface of the planet respond to climate modifications?

The marine environment has responded to this warming. We hear a lot these days about the shrinking arctic sea ice pack. In September 2007 it was 38 percent below average (1979–2000),[1] and the decline appears to be more severe than many of the models projected. The decline in sea ice has been possible to measure because sea ice is visible from space. When sea ice melts, the surface of the planet changes from ice to water, and from white to black. This change in color allows the entire arctic ice pack to be assessed spatially over the period of satellite record.

Terrestrial changes require more effort to detect and evaluate, but they do exist. In this book I will present evidence for changes in arctic vegetation, glaciers, and frozen ground. Most of the evidence presented comes from repeated photos, as described in the Preface. The repeated photos contain evidence for change that is hard to dispute, but the attribution of the changes to warming is not so straightforward.

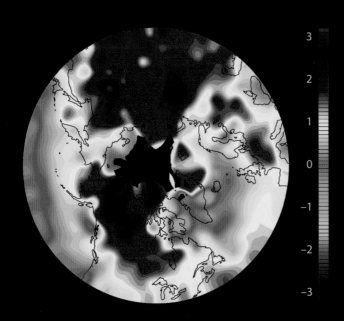

Figure 2   Surface air temperature change (°C) in northern regions during the period 1959 to 2008, relative to the mean. WILLIAM CHAPMAN AND JOHN WALSH

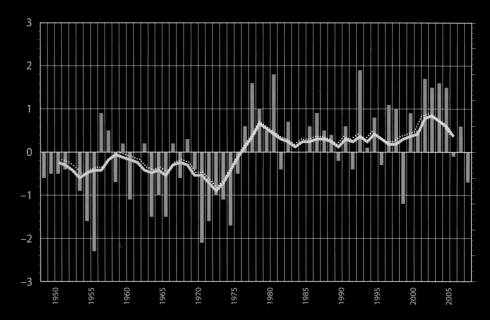

Figure 3   Alaska air temperature anomalies (°C) since 1949, relative to the mean. This trend is derived from climate stations all over Alaska, and the overall increase in temperature has been 1.9°C. The warming is exaggerated in the Alaskan arctic, where the lone station in Barrow recorded a 2.3°C increase. ALASKA CLIMATE RESEARCH

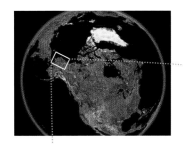

The Pristine Arctic Landscape

THE ARCTIC LANDSCAPE OF northern Alaska consists of low-growing vegetation largely devoid of trees and of human presence. Although the southern boundary to this region is sometimes considered to be the Arctic Circle, the southern boundary is, for our purposes, the northern extent of the boreal forest. This boundary lies along the axis of the Brooks Range, an east-west trending region of mountains and glaciers that is 150 km wide and 600 km long (**Figure 4**). North of the Brooks Range are the North Slope uplands, a gently undulating tundra punctuated by north-flowing rivers that originate in the Brooks Range (**Figures 5 and 6**). Farther north, the uplands give way to the coastal plain, which stretches to the arctic coast (**Figure 7**). Much of the Iñupiat population lives along the coast, in scattered villages such as Barrow, Kaktovik, Point Lay, and Kotzebue (**Figure 8**).

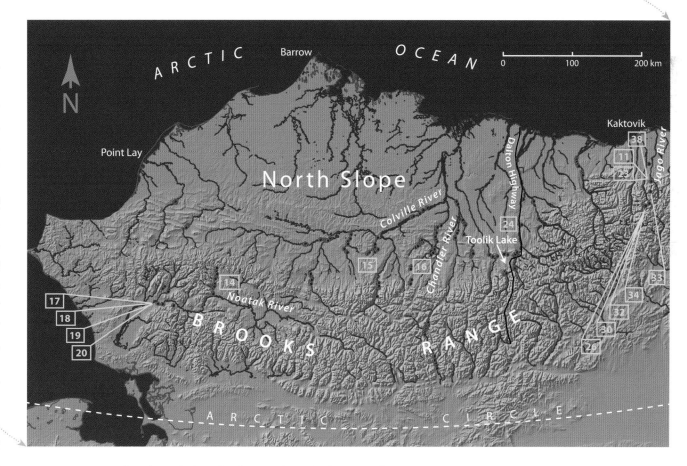

Figure 4 Map of northern Alaska. Yellow boxes indicate the locations of repeat photography figures in this book, and red dots indicate locations of vegetation repeat photography described in the text.
GLOBE INSET: GOOGLE EARTH. BACKGROUND SHADED RELIEF: USGS

The Alaska Arctic has been occupied by Iñupiaq people for more than ten thousand years (**Figure 9**). At about the same time that these people were coming to Alaska—via the Bering Land Bridge from Siberia—the glacial age was waning and leaving a dry grassland. Over several thousand years, the grassland was transformed into a tundra similar to the modern one. **Figure 10** is a simplified pollen record from a lake sediment core, and it shows how vegetation assemblages and the modern tundra have evolved since the waning of the glacial age. Before that, glaciers extended north of the mountains onto the rolling uplands, where they have left behind moraines now covered in tundra vegetation. It was one such moraine that Ed Sable and George Kunkle identified and mapped along the Jago River (in the Arctic National Wildlife Refuge) in 1957, and that will gain our attention later (back cover photo and **Figure 11**).

Nowadays, the mean July air temperature for northern Alaska is 11°C (52°F), and the mean January air temperature is –18°C (–28°F). Snow is typically on the ground from the end of September to early May, although extreme wind events can scour the landscape at any time of the year, leaving bare tundra exposed to the cold temperatures. The sun doesn't rise from late November to late January and doesn't set from late May to late July. Caribou and birds numbering in the hundreds of thousands migrate to the region. Dall's sheep, musk oxen, bears, wolves, and moose are some of the larger permanent residents (**Figure 12**).

Figure 5 (LEFT) Late August colors in the tundra of the Ivishak River headwaters. The view is looking south from the North Slope to the Brooks Range.

Figure 6 (ABOVE) Brent Sass and Danny Dominick leading dog teams down the frozen Nanushuk River, Brooks Range, late April.

5

Figure 7 (LEFT) Arctic coast near Barrow, in March. The aurora is a backdrop for the sea ice blocks and bowhead whale skull.

Figure 8 (RIGHT) Greta Myerchin and Lisa Garrison next to drying meat along the flat coastline in the village of Kotzebue, in northwestern Alaska.

Figure 9 (BELOW) Ancient rock cache late in the evening (left) and under the midnight sun (right). The cache is located on a promontory in the western Noatak. From here, views for ancient hunters would have been exceptional. It is thought that ancient people came to this region about 11,000 years ago.

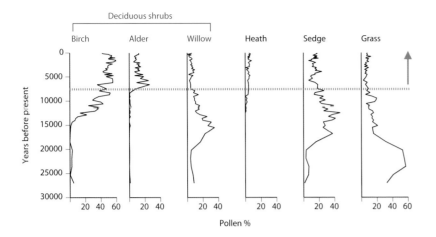

Figure 10 (ABOVE) Pollen record from a North Slope lake core going back 30,000 years. The vertical axis is time, with the present on top. The horizontal axes represent the relative abundances of pollen. Plant abundances, as measured by the pollen record, have been relatively stable for the last 7,000 years (arrow). When assessing the changes in the repeat photography presented in this book, you should consider whether the changes observed in the photos constitute only another blip of the limited variability seen in the last 7,000 years, or whether the changes in the photos might be indicative of larger shifts, such as those denoted by the dotted line. W. WYATT OSWALD

Figure 11 (RIGHT) Repeat photo pair from the Jago River valley. The old photo (top), from the Sable collection, is from 1957 (George Kunkle), and the new photo (bottom) is from July 29, 2007. The camera positions are nearly identical in the two photos, but in the new photo willow shrubs have replaced alpine flora and are obscuring the ponds and many boulders seen in the old photo. Other boulders have migrated downslope.

Figure 12 (BELOW) Brooks Range caribou migration in November. Musk oxen (LEFT) and moose (RIGHT) in fall foliage.

Vegetation

OUTSIDE OF THE RIVER floodplains, the tundra ranges from wet to dry, and from rocky and devoid of vegetation on ridges and in the mountains to wet and crammed with tussocks on the coastal plain. Tussocks are ankle-breaking protrusions that resemble mannequin heads with grass (or sedge) wigs. Where there is an organic layer, it insulates and is underlain by frozen ground within a meter of the surface. At 29°C (85°F) in July you can stick a steel rod into the ground and hit an impenetrable, impermeable surface a half-meter down. The organic layer at the top of the soil profile is often peat, which is an accumulation of organic material resulting from aboveground biological production being greater than belowground decomposition. Insulated cold soils and often anoxic conditions slow decomposition, which further promotes the accumulation of organic material. Smaller shrubs (less than a meter high) are nearly ubiquitous. Larger shrubs (between one and four meters high) grow on slopes or in the floodplains (**Figure 13**), probably because both places enjoy good drainage and a flux of nutrient-rich water.

Elsewhere on the planet, landscapes experience periodic large-scale disturbances, such as hurricanes or fires, that impact vegetation. After a fire in the boreal forest, for example, the vegetation begins a natural cycle called

Figure 13 Midnight sun shining on the Nimiuktuk River, western Brooks Range. The dark shrubs are mostly alders 1–4 m tall (*Alnus viridis* ssp. *fruticosa*), and the lighter shrubs are shorter species of willow (*Salix* spp.) and birch (*Betula* spp.). Poplar (*Populus balsamifera*) tree stands can be seen to the left of the river channel. Our tents are colored dots on the sandbar.

Figure 14 Repeat photo pair showing shrub expansion on the Nimiuktuk River, western Brooks Range. The old photo (above), from the Col photo collection, is from 1950, and the new photo (right) is from 2002. The white patch in the background of both photos is remnant aufeis (frozen overflow) from the previous winter.

succession, starting with the first plants to take root after the fire and ending with the mature forest that we are familiar with; when fire returns, the cycle repeats. Because of these disturbances and consequent cycles in vegetation, it is difficult to make observations of vegetation over time and then to attribute any changes to changes in climate. The attribution to climate is not impossible, but it would require a large number of carefully chosen observations to distinguish any changes due to climate from changes due to succession.

In the Arctic, however, fire and hence succession are rare beyond the reach of the floodplains, so we need not worry about what portion of the vegetation cycle we are witnessing. The Arctic is thus a natural laboratory for studying landscape responses to changes in climate, because the vegetation exists more or less in equilibrium with climate. The equilibrium, however, is probably complex and less than perfect, with time lags between changes in climate and the terrestrial response.

Evidence for Shrub Expansion

In 1999, snow physicist Matthew Sturm caught wind of some old aerial photos in Anchorage that were reportedly headed for the dump, and he

insisted that one box be shipped to Fairbanks for a look. After a preview, Sturm recognized their potential importance and had the five-thousand-print Col photo archive shipped to Fairbanks over several months. The photos were 18 x 9 inch contact prints taken from 1946 to 1952 as part of the geologic reconnaissance in the Naval Petroleum Reserve No. 4, now the National Petroleum Reserve-Alaska. The low-altitude oblique photos covered every major drainage on the North Slope, as well as many drainages in the Brooks Range. Because these prints were made from 18 x 9 inch negatives, the resolution of the photos is exceptional.

Sturm, at the Cold Regions Research and Engineering Laboratory (CRREL) in Fairbanks, Alaska, contacted ecologist Charles Racine at CRREL in Hanover, New Hampshire, and together they repeated sixty-seven

Figure 15 Repeat photo pair showing shrub expansion on the Oolamnagavik River, central North Slope. The old photo (left), from the Col photo collection, is from August 11,1948, and the new photo (right) is from July 27, 2002. On the floodplain the alder expansion is obvious, and on the slope the expansion is more difficult to see. The old photo has sustained some water damage.

Figure 16 Repeat photo pair showing shrub expansion on the Chandler River, central North Slope. The old photo (left), from the Col photo collection, is from July 4, 1948, and the new photo (right) is from July 28, 2001. Numbers indicate the same rocks in the old and new photos.

of the old photographs to see what they would find. This required photo-grammetry to triangulate the location of the old photos, which could then be relocated by helicopter and replicated. In short, they found more shrubs in the new photographs than in the old photographs. Sturm and I subsequently spent the following three summers rephotographing more than three hundred landscapes, covering sections of twenty rivers, to confirm and expand on this result.

The primary change in the landscape was an increase in the coverage of green alder shrubs (*Alnus viridis* ssp. *fruticosa*), willow shrubs (*Salix* spp.), and birch shrubs (*Betula* spp.). Most shrub patches in the photographs have expanded (1) by expanding their previous patch boundaries, (2) by becoming more densely shrubby within the patch boundaries, or (3) by having their constituent individuals grow larger. Shrub patches that had changed generally exhibited all three types of shrub expansion (**Figures 14–16**).

Most of our attention was originally focused on the valley slopes leading down to the rivers, because the slopes often hosted large shrubs visible in the old and new photographs, and because we considered these surfaces stable and free from periodic disturbances like flooding or fire. We applied a gridding technique to quantify the changes in shrub cover from the old photos to the new photos, and we found that shrub cover on valley slopes has increased from 15 to 20 percent. Of the 1,335 grid cells analyzed, 894 registered an increase in shrubs, 428 registered no change, and 13 registered a decrease. Of the 155 repeated photos analyzed quantitatively, 135 showed an increase in shrubs, 20 showed no change, and none showed a decrease.

After extracting this quantitative information from what we considered to be stable landscape positions, we shifted our attention to the river terraces and floodplains, where we would be dealing with noise in the signal due to periodic flooding. River terraces are elevated above the active floodplain, and they typically have a continuous organic layer of vegetation underlain by soils and aggrading permafrost. The terraces have experienced an explosive increase in shrubs since the original photos, often in the form of large new areas being colonized by shrubs. Of the seventy-two photo pairs analyzed for changes on river terraces, forty-seven showed an increase in shrub cover, twenty-five showed no change in shrub cover, and none showed a decrease.

We found similar trends in the floodplains. Of the forty-nine photo pairs analyzed for changes in floodplains, thirty-eight showed an increase, eight showed no change, and three showed a decrease. The trends, however, contained more noise than had those from the adjacent slopes, thus making interpretation of the results more difficult. Initially we thought that the noise might be due to flooding; we would naturally expect to see changes in the floodplain, with sandbars stabilizing in some areas and hence promoting shrub growth and shrubs being inundated in others. But the increases and decreases in shrub cover should then be roughly the same. Instead, river channels in the new photos appear narrower, as if constrained by vegetation. Meanwhile, vegetation is clearly colonizing and stabilizing formerly active portions of the floodplain. But it is unknown whether the changes in floodplain vegetation and morphology are simply a continuation of the shrub expansion occurring outside of the floodplain or whether the floodplain changes are, in part, being driven by changes in river discharge.

Getting Our Feet on the Ground

Since repeating the Col photos, I have been on the lookout for old landscape photos that would be suitable for repetition, especially photos taken from the ground during early expeditions in northern Alaska. Whenever my travels take me near Denver, I make a point to visit and search the USGS archive for vegetation photos in regions I might someday visit.

The early geologic expeditions followed river corridors, occasionally encountering local inhabitants who had occupied the region for thousands of years. The geologists were usually photographing broad landscapes, rock exposures, or the local people. Their photos are often rich with Iñupiaq culture and wilderness, but the photographers were not working with my purpose in mind. That is to say, they were not worrying about whether the exact location of a photo could be rediscovered decades later or whether the photo showed interesting vegetation. So most of the photos in the collections have little value for detecting vegetation change. With all the excitement of the chase, searching for useful photos is a lot of fun, but the odds of success are long.

Several elements must be present in a photo if it is to have potential for repetition. First, the photo must have proper exposure and focus, neither of which was trivial in the early days of photography. For our purposes the photo must also contain some vegetation not on a floodplain. And the photo must have enough distinct foreground and farground to permit triangulation. The spoiler for many potential repeat photos is the lack of identifiable foreground, particularly in the nondescript tundra, making the photo impossible to relocate precisely.

I was therefore excited to find a number of promising photos from Philip Smith's USGS party, who descended Noatak Canyon a century ago. I then enlisted Google Earth and a special mouse that allows you to fly around, and I toured the virtual Noatak Canyon until I had recreated the views seen in Smith's photos. After recording GPS locations for each photo, it was a matter of boating the river, traversing the canyon terrain, and finding the simplest access to the locations.

The twenty-three repeated Smith photos from 1911 collectively confirm that shrubs increased along the Noatak during the twentieth century (**Figures 17–20**). In addition, the vegetation has encroached on the narrowed Noatak River floodplain, and there are more poplar and spruce trees both on developing river terraces and on upland sites. The dramatic treeline spruce expansion documented in repeated photos both from here (**Figure 20**) and from a nearby tributary (the Kugururok River) provide an interesting contrast to the general boreal forest decline in southerly regions.

The Verdict from Satellites

Satellites have been orbiting and photographing Earth since the 1970s. Comparison of recent images with those from decades earlier reveals an apparent "greening" across the Arctic (**Figure 22**). Unlike disappearing sea ice, though, which shows up as a marked change from light to dark on a pair of old and new satellite images, the change from old to new satellite images of vegetation is rather subtle. A satellite image of tundra, containing thousands of repeated pixels, changes only from green to greener. Nevertheless, in the last several decades greening has been documented over large parts of the Arctic.

Figure 17 Repeat photo pair showing shrub expansion along Noatak Canyon. The old photo (top), from the Philip Smith USGS collection, is from August 16, 1911, and the new photo (bottom) is from July 4, 2008. Following a bushwhack along the ridge of the canyon, I was able to locate the knob in the bottom center of the old photo; together with more distant features, the knob was ideal for triangulation. The ellipse indicates an area where shrub patches have expanded.

Technicalities of Greening

The "greenness" is measured using a vegetation index called Normalized Difference Vegetation Index (NDVI). NDVI combines signals from the red (wavelength; $\alpha R = 0.58 - 0.68\,\mu m$) and near-infrared ($\alpha NIR = 0.72 - 1.10\,\mu m$) satellite channels ($NDVI = [\alpha NIR - \alpha R] / [\alpha NIR + \alpha R]$). Deciduous leaves are composed of chlorophyll, which makes them green. Green appears dark when viewed through the red channel. On the other hand, green appears bright when viewed through the near-infrared channel, due to the presence of chlorophyll. So the greenness is measured using NDVI, which correlates well with photosynthetic activity and biological production. Does this greening in the satellite images represent a real trend in vegetation change, or could it somehow be an artifact of the satellite sensor?

Satellite sensors deployed in the 1970s and 1980s were not built for monitoring long-term trends. Because the satellites drift subtly over time, the record often includes several different replacement sensors, ostensibly replicas of the original sensor. Also, there are different methods to correct for atmospheric distortions and to reduce the data. Because of these complications, it was hard to know whether or not the satellite trends represented an actual change in vegetation.

However, in the decade following the early papers on NDVI trends, multiple satellite sensors and multiple data reduction algorithms produced the same trends. Now, numerous studies document the greening, and some of these have used ground-truth data to show that the greening is caused by increases in deciduous shrubs. Other greening studies show that the growing season has lengthened, particularly in the spring, such that "green-up" now occurs earlier than before.

The repeat photography and repeated satellite images together tell a consistent story of widespread increasing shrubs in northern Alaska. The aerial repeat photography is most effective for detecting changes in shrubs taller than half a meter. The satellite imagery is most effective for shrubs smaller than about a meter (larger shrubs saturate the NDVI signal). Repeat photography covers primarily river valleys and floodplains, and the satellite images cover the entire landscape, including broad tundra benches with scattered small shrubs. The repeat photography and satellite trends are therefore complementary in their spatial coverage and size of shrubs assessed. Both point in the direction of increasing deciduous shrubs.

Figure 18 Repeat photo pair showing shrub expansion at the entrance to Noatak Canyon. The old photo (left), from the Philip Smith USGS collection, is from August 18, 1911, and the new photo (right) is from July 4, 2008.

Figure 19 Repeat photo pair showing shrub expansion near Noatak Canyon. The old photo (top), from the Philip Smith USGS collection, is from August 16, 1911, and the new photo (bottom) is from July 4, 2008. Pinpointing the old camera location required a lot of back- and-forth searching, but eventually the river and lake produced what seemed to be an acceptable triangulation. It wasn't until later that close examination of matching tussock orientations in the old and new photos

revealed that the new photo was taken within two meters of the old. In the new photo there are two to threefold more alder shrubs, the lichens (white specks in old photo) have been replaced by grasses, and the spruce trees and shrubs are climbing up the left slope. A study using historical vegetation data from the same region recently showed a similar shift from lichens to grasses and shrubs since 1981. The bright scar and dark spot on the old photo are technical flaws.

George Gryc

George Gryc at Umiat, May 25, 1947, and in 1995. PROJECT JUKEBOX

From interview with the author on December 11, 2004 and with Karen Brewster, Project Jukebox, on September 21, 2004:

George Gryc was a PhD candidate in geology at the University of Minnesota when World War II began. He had the choice to enlist in the military or to come to Alaska as part of the war minerals program. In 1944 he chose the latter. As USGS employees responsible for exploring the potential for resource development in what later became the National Petroleum Reserve–Alaska, he and his colleague Bob Coates were dropped off by legendary bush pilot Sigurd Wien on a sandbar of Prince Creek near the Colville River. They traversed the Colville River to the Umiat anticline, where they mapped the folded rock and oil seepages that had been seen from the air during flight reconnaissance.

The following year, Gryc led one of several mapping parties descending onto rivers with headwaters in the Brooks Range. In the spring of 1945, Gryc's USGS party spent four weeks on the eastern edge of Chandler Lake mapping bedrock, making preparations for the float trip, and waiting for the ice on the river to go out. There were about fifteen resident Nunamiut (inland Eskimo) people in the Chandler Lake area, and they were led by Simon Paneak and Elizah Kakinya. The group lived in willow and skin huts at the south end of the lake and subsisted on fish and caribou in the traditional way.

Gryc and the cook, Charles Metzger, who later wrote about the trip, described how Paneak or Kakinya, neither of whom spoke much English at the time, would show up in the late afternoon at the USGS camp and stay for dinner. Gryc and the USGS party would eventually do the same at the Nunamiut encampment. As the groups got to know each other better, they exchanged goods and skills, the latter being mostly Nunamiut men demonstrating their ability to hunt, mush dogs, kayak, and ice fish. In exchange for dehydrated vegetables, powdered milk, a hundred feet of rope, and whatever the USGS party couldn't eventually take downriver, the Natives agreed to use their dog teams to sled about two thousand pounds of USGS party gear to a point twenty-five miles downriver (north), below the worst of the rapids. It was a steal for both sides.

So, after the snow melted but before the ice was out on the river, all the Nunamiut people and their dogs arrived to haul the extra freight and geologists downriver. It was an opportunity for the geologists to see the still-frozen rapids ahead of them, and for the Natives to show the geologists the location of the "rocks that burn." En route, the Natives allowed the white guys to have a turn at mushing dogs.

The Nunamiut often listened to the radio. As Gryc recounts, they frequently heard of a man named Roosevelt (FDR), and they decided that he was a good guy. One day they stopped hearing about Roosevelt on the radio, and it was not long after that Paneak named his child Roosevelt. In 2000 at the Toolik Field Station, I narrowly missed Roosevelt Paneak's stop there, because at the time I was unaware of his link to history.

Roosevelt was not the last of Simon Paneak's offspring, and in fact Simon's wife, Susy, was pregnant when George Gryc's USGS party was at Chandler Lake. Not long after the party departed, Susy Paneak gave birth to her son George.

The two parties—the Nunamiut and the USGS group—were born of completely different eras. Each was remarkably competent in its own domain but somewhat inept in the other's. Yet the extent to which they got along, sharing their skills and goods that spring, remains a testament to the potential for amicable relations across other cultural chasms.

At the time, Sigurd Wien periodically landed on the frozen Chandler Lake to provide goods to the Nunamiut in exchange for furs. At one point, Wien noted that he could make more frequent visits if the Natives moved to the broad pass 30 km to the east. Paneak, Kakinya, and the others agreed, and in 1947, they moved to the Anaktuvuk Pass area. In 1949 they were joined by another group from the Killik area, and the settlement marked the end of the purely nomadic existence in northern Alaska. Gryc went on to become the head of the Alaska Branch of USGS, and eventually the Pacific Branch. Over the years he returned several times to the village of Anaktuvuk Pass to see his old friends.

George Gryc passed away in April 2008. In Anaktuvuk Pass, George Paneak was the mayor until he passed away in September 2009.

George Gryc (front right), Ernest Lathram (front left), James Zumberge (back left), and an unidentified field assistant pulling one of their boats over the frozen tundra to begin their descent of the Sagavanirktok River, 1946.
IMAGE COURTESY OF STEPHEN GRYC

Simon Paneak constructs a willow and caribou skin hut for research purposes in Fairbanks, decades after his interaction with Gryc's party at Chandler Lake.
UNIVERSITY OF ALASKA MUSEUM OF THE NORTH

Shrub Expansion and Warming

The vegetation of arctic Alaska is becoming shrubbier. It is plausible to attribute the change to a warmer climate. What is the evidence for such a connection?

At the Toolik Field Station just north of the Brooks Range, ecologists have been manipulating the tundra and watching it respond under a variety of different warming and fertilization scenarios. The goal is to predict how arctic vegetation might respond to warming and also to understand what factors might limit growth in the Arctic, whether the factors are physical limitations such as temperature or nutrient limitations such as nitrogen or phosphorous. At Toolik, ecologists Terry Chapin, Gus Shaver, and many others have performed these manipulations since the early 1980s. They fertilized some vegetation plots with nitrogen or phosphorous; in others they reduced the light, increased temperature, or added snow. The vegetation generally responded to artificial fertilization and elevated temperatures by favoring shrub growth.[3]

Differences in the responses between various treatments have shed light on the specific mechanisms driving the changes. The results from Toolik suggest that shrubs would do better under a warmer climate.

Another suggestion that the shrub expansion is linked to warming comes from examination of the active (thaw) layer in shrub patches. In 2008 we visited twenty-six shrub patches across northern Alaska that we had identified as expanding or stagnant on the basis of repeat photography. We found that, on average, the active layer was significantly deeper in the sixteen expanding patches than in the ten stagnant patches. This suggests that the soil temperatures had been higher in the expanding patches.

There is another, admittedly somewhat soft, argument for associating the shrub expansion with warming. Change in a single location—be it in a plot, a photo, or a pixel—might be attributed to chance or to unknown features peculiar to the particular location. Arctic vegetation, however, is changing on a broad scale. The widespread nature of the change suggests that, whatever factor is

responsible for the change, it is operating on a similarly large scale. It is hard—though not impossible—to imagine what that factor might be, if not warming.

But in fact that factor might conceivably be a change in precipitation—either rain or snow. Until recently, however, weather stations have only been sparsely scattered across the Arctic, so there are few historical precipitation records from which to infer trends. It is unknown, for example, whether or not August rainstorms have become more severe. Such a change could impact vegetation.

Figure 21 (RIGHT) Ricky Ashby in his homemade chair, at his remote Noatak River cabin. The repeat photo of Figure 20 was coincidentally just upriver from his cabin, so he wandered up the bank to investigate and eventually invited us to spend the night. The wisdom inherent in his traditional subsistence lifestyle was an inspiration to our party.

Figure 20 (BELOW) Repeat photo pair west of Noatak Canyon. The old photos (top), from the Philip Smith USGS collection, are from August 18, 1911, and the new, panoramic photo (bottom) is from July 5, 2008. The new photo has more shrubs, more spruce trees (arrow), and recently formed river terraces.

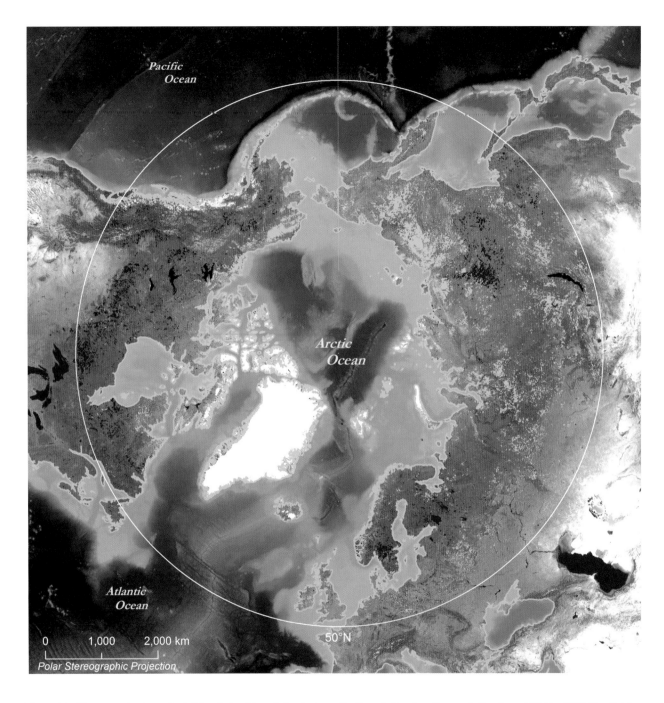

Figure 22 Circumarctic view of satellite-derived trends in photosynthetic activity of arctic vegetation from 1981 to 2005. The light green represents areas that have "greened" since 1982, a change that scientists largely attribute to an increase in deciduous shrubs and perhaps grasses. The red areas represent concurrent declines in the boreal forest.[4] SCOTT GOETZ

Snow is critical for protecting vegetation and for insulating the ground. But it is highly variable in space and time. The redistribution of snow by the wind creates a blanket of highly variable thickness. The depth can vary by a factor of three from one year to the next, and the blanket can arrive early or late in the winter.

Because of its variability, snow depth has proven difficult to measure over large areas using satellites. The only reliable way to measure snow depth is to visit the location at the end of the winter (May) and make thousands of measurements over the area of interest. This is usually not feasible over large areas. Even if you could measure the late-winter snow depth, you would not know the snow depth in January of that winter, nor would you know how deep it was the year before. It is thus unknown, for example, whether or not snow cover has become deeper over the years. Again, such a change would impact vegetation.

The evidence for expanding shrubs in arctic Alaska is clear. The evidence linking that expansion with warming is good but not unassailable.

The Face of Change

George Gryc and a small party of USGS geologists and field hands descended the Chandler River (shown in **Figure 16**) in 1945. For navigating some sections of the river they referred to crude trimetrogon reconnaissance photography, but for other sections the expedition scientists often had no idea what lay around the bend. The cook, Charles R. Metzger, would later become an English professor and recount the group's experiences with Nunamiut people at Chandler Lake, as well as the ensuing epic float trip, in his book *The Silent River*.

It was an honor, then, for me to retrace some of that trip a half century later. Together with Matthew Sturm and Charles Racine, I floated the lower, milder sections of the Chandler and Ayiyak rivers in 2000 and 2001, to sample and characterize the vegetation. The landscape turned out to be unlike that found elsewhere on the North Slope. Shrubs were as large and dense as any we had seen; we were often bushwhacking through waist-high birch shrubs and much taller alders and willows. Poplar trees and moose sightings—the latter an indication of abundant shrub forage—were commonplace.

Several years later, on a hunch that snow or growing season length might be responsible for conditions favoring shrub growth on the Chandler, I went

Figure 23 Repeat photo pair from Jago Lake area, eastern North Slope, showing relatively little change in the landscape. The old photo (top), from Ed Sable's collection, was taken in 1957 by George Kunkle. The new photo (bottom) is from July 29, 2007.

airborne in early May, with lakes frozen over and snow still on the ground, and repeated photos on the Colville and Chandler rivers. While the Colville River and most of its northern tributaries were covered in snow and the river ice was not yet out, the Chandler River valley was mostly snow-free, and only shelf ice remained on the open, navigable river. Perhaps it was ground temperature and season length, driven by microclimate or groundwater inputs, or perhaps it was something else, but the vegetation on the Chandler, and neighboring Ayiyak and Siksikpuk rivers, appears to be a step ahead of vegetation on rivers elsewhere on the North Slope. The shrubs in these three valleys are larger and more abundant than anywhere on the North Slope, and it gives us some idea of what other parts of the region might soon look like, if current trends continue.

Thus the Chandler River valley hints at an answer to the important question: What does an arctic landscape look like that has responded to increases in temperature?

It's Not All Changing

In 2007, glaciologist Matt Nolan and I were on the way back to our Jago Lake camp under a darkening sky. Armed with an old George Kunkle photo that showed Ed Sable among some "frost boils" somewhere near here, I hurried ahead along an ancient (more than 20,000 years old) moraine,[5] searching for the frost boils. I had known when I scanned the photo from Ed Sable's 1957 field notes that relocating it was a long shot, but the mountaintops in the photo provided some hope for triangulation. Those mountains, though, were presently obscured by cloud, and so the odds of success were even longer. I criss-crossed the ridge for a while, looking for a certain large rock and for the same arrangement of gravel and vegetation seen in the old photo. And then there it was! After a few photos, I asked Nolan if he would strike Sable's pose from the old photograph. Nolan protested that his knees didn't bend that way, but he nevertheless assumed a position comparable to Sable's. As Nolan reached down to his right, he realized that in the photo from 1957, Sable had been picking crowberries (*Empetrum nigrum*).

Figure 23, containing both Kunkle's photo and the repeat, shows that this location has not visibly changed in a half century. The landscape looks exactly the same, from the low vegetation right down to the orientation of individual rocks. There are no large shrubs in the old photo and no large shrubs in the

new photo. Remarkably, this location is less than a mile from that of **Figure 11**, which shows dramatic shrub expansion. There, the increase in shrubs probably reflects some change in the substrate—perhaps warmer, drier, or more nutrient-rich soils. The lack of visible change in **Figure 23**, however, does not necessarily mean that nothing is changing in the soils. The soils may well be responding to the elevated temperatures, but there is little vegetation present to express that change. This location, and similar ones across the Arctic, are not likely to host large shrubs in the immediate future.

Figure 24 is another relatively unchanged landscape. The individual alder shrubs visible in the new photograph are, in most cases, the exact same individuals present in the old photograph. But wait—if the shrubs looked exactly the same in 1949 as they do now, then is there any reason to believe that they didn't look exactly the same in 1900, 1850, and 1800? In all likelihood, these alder shrubs—and it is probably true for willow and birch as well—are hundreds of years old! Out of curiosity, we sampled multiple large stems from several individuals, and we rarely found stems older than fifty years, and never enough to support the shrub that was clearly in that location fifty years ago. Apparently, old stems eventually die, are buried, and then decompose beneath the shrub, while new stems grow from their remains. Despite the ongoing turnover of the individual stems, these individual shrubs likely persist for hundreds of years.

The unchanged photographs, though they constitute a small minority of my repeat collection, are my favorites. I like them because they give, if incorrectly, a reassuring sense of timelessness to pristine arctic landscapes. At their best, these landscapes are untouched by mankind and untouched by the catastrophic forces of flooding and fire. They, too, may eventually bend under the pressures of a changing climate, but for now, they are resisting.

Figure 24 (OPPOSITE) Repeat photo pair of shrubs on the Lupine River, a tributary of the Sagavanirktok River, North Slope. The old photo (top), from the Col photo collection, is from 1949, and the new photo (bottom) is from July 28, 2001. This shrub patch remains almost exactly as it was a half-century earlier. The numbers mark ten individual shrubs that are the same size in the old and new photos. The stability in this shrub patch suggests that individuals are hundreds of years old.

Glaciers

Leffingwell's Legacy

Ernest K. Leffingwell spent nine summers and six winters on the arctic coast between 1906 and 1914 (**Figure 25**). He sometimes enlisted Natives to assist in his dog-mushing, boating, and packing expeditions along the coast and into the eastern Brooks Range, but many of his trips were solo. In 1907, during one of his thirty-one trips through the region, Leffingwell ventured into the mountains and photographed the country he beheld. He ventured into the heart of the eastern Brooks Range and photographed Okpilak Glacier from a terrace overlooking the terminus. He correctly estimated the mass of ice to be the largest glacier in the region.

Upon returning to Washington, D.C., Leffingwell was granted office space at the USGS by Alfred H. Brooks, after whom the Brooks Range was named. Leffingwell's contribution to science and methods of remote arctic travel is meticulously archived in his USGS Professional Paper 109.[6] Brooks himself writes in the Preface to 109 that, prior to Leffingwell, "The region as a whole . . . presented an almost complete hiatus in the scientific knowledge of Alaska, and Mr. Leffingwell has performed a most valuable service in mapping its geography and geology." Leffingwell was mapping bedrock in a remote corner of Alaska and doing so independently of USGS or other agency funding.

In 2007, glaciologist Matt Nolan and I returned to the spot of Leffingwell's panoramic photos of Okpilak Glacier. The landscape was hardly recognizable from the old photos. The prominent feature from the 1907 photograph—the

mass of ice constituting the terminus of the Okpilak Glacier—was miles up-valley and looked a bit pathetic (**Figure 26**). The repeated panoramic photo spoke for itself, suggesting that radical changes were underway in Brooks Range glaciers.

It's Not Breaking News

Ed Sable had reached the same conclusion a half-century earlier. Sable had come to Alaska as a field assistant for a USGS party in 1946. Over the following fourteen years, Sable visited almost every corner of the Brooks Range and North Slope, mapping the bedrock geology and looking for natural resources.

In 2003, at the age of eighty, Sable was cleaning out his office at the Denver USGS and was wondering what to do with the aerial reconnaissance photography that he had used to complete his masters and dissertation work in the late 1950s and early 1960s. He called his old supervisor at the USGS, George Gryc, for suggestions. By that time I had interviewed Gryc about his own experiences in northern Alaska, and Gryc knew that I was looking for old landscape photos. Gryc kindly asked Sable to send the photos my way,

Figure 25 (ABOVE) Ernest Leffingwell with his sled dogs at his dwelling on Flaxman Island, arctic coast, 1909.

Figure 26 (RIGHT) Repeat photo pair from the Okpilak River valley. The old photos (top, Ernest Leffingwell, Matt Nolan stitching) are from 1907, and the new panoramic photo (bottom, Matt Nolan) is from August 6, 2007.

and within several weeks I was opening a box with a set of 18 x 9 inch contact prints, showing rugged mountains and valley glaciers—in locations that had been missing from our Col photo collection. The photos contained dozens of glaciers in the Okpilak Valley, including the Okpilak Glacier itself, and many tributary glaciers. The photos had a fine-art—almost painted—quality, and were overlain with Sable's blue crayon outlines of geologic strata, glacial features, and deformation history, all annotated with geologic symbols.

I was pretty excited. After hearing my reaction, Sable shipped more aerial photos from the same era; these were vertical and trimetrogon geologic reconnaissance photos that he had used to spatially connect his field observations. Sable also shipped a binder with mounted photos from his glacier fieldwork in 1957–1958, some of which are repeated here. Finally, in a later shipment, there was a reprint of his dissertation mapping the bedrock of the Romanzof Massif, published as USGS Professional Paper 897 (1977), and a much thinner reprint "Recent Recession and Thinning of Okpilak Glacier, Northeastern Alaska," (1961) from the journal *Arctic*.[7]

As a student in the 1940s and 1950s, Sable knew Leffingwell's Professional Paper 109 very well, as it was still an authoritative source for geology of northeastern Alaska. Sable, who by 1958 had a geologic background but no formal education in glaciology, had spent twelve summers in northern Alaska developing a keen understanding of glacial deposits and glacier dynamics. In 1958, he repeated some of Leffingwell's 1907 photographs of Okpilak and nearby glaciers to show that they were shrinking. Sable did not carry Leffingwell's photos into the field, but he remembered from one photo the orientation of the hanging glacier across the valley to the terminus of the Okpilak Glacier in the foreground. When he tried to repeat that photo, he was unable to align the Okpilak glacier terminus with the hanging glacier behind, because the terminus had retreated so far up the valley. The photos confirmed Sable's inferences drawn from glacial deposits from numerous Brooks Range glaciers. In his paper, Sable writes,

> All glaciers observed in the Romanzof Mountains that are comparable in size to the Okpilak Glacier show evidences of recent retreat and thinning similar to those described above. Present termini of glaciers have commonly withdrawn several hundred feet leaving unweathered, unstable terminal moraines, till ridges, or formless drift at many localities. The

writer believes that the shrinkage of these glaciers coincided with that of the Okpilak Glacier. It seems likely that climatic conditions leading to a general shrinkage of valley glaciers in the Romanzof Mountains and perhaps in the entire Brooks Range have occurred during the last half of the 19th and the first half of the 20th century.

Sable was a half-century ahead of his time.

Glaciers have since become the poster child for terrestrial changes, because their dramatic retreats are easily identified using repeated photos, and because the relationship between temperature and ice is more plausible (perhaps deceptively so) than that between temperature and vegetation. However, very few repeat glacier photos are of arctic glaciers, largely because of the difficulties in getting there, especially fifty to a hundred years ago. The beauty of the Brooks Range glaciers for inferring climatic change is that they are simpler than their larger counterparts to the south that calve into the ocean or periodically surge. This means the mass balance and ice dynamics of these glaciers are more representative of regional climate trends.

In 2006, Anthony Arendt completed his dissertation at the University of Alaska Fairbanks, where he used airborne laser altimetry to show that glaciers in all parts of northwestern North America are retreating.[8] Arendt synthesized and expanded upon work by glaciologists Keith Echelmeyer and others, and calculated the loss of glacier ice over eighty-six Alaska, British Columbia, and Yukon Territory glaciers, including eleven Brooks Range glaciers. Arendt found that all eleven of the Brooks Range glaciers were smaller than they had been in the 1970s.

McCall Glacier

In August 2007, a group of us led by Matt Nolan trekked into the eastern Brooks Range, from the coastal plain to the high glaciers of the Arctic National Wildlife Refuge, or ANWR, and eventually back to the coastal plain. It was a multifaceted expedition designed to repeat glacier and vegetation photos, to continue ongoing measurements on McCall Glacier, and to collect baseline vegetation information for future monitoring purposes. Nolan had tracked down Austin Post's collection of vintage glacier photographs and was interested in documenting not only the retreat of McCall Glacier but also the effect of the glacier on the tundra vegetation along McCall Creek and the Jago River.

Ed Sable

From interview with the author on July 6, 2007:

Ed Sable was born in 1924 in Rockford, Illinois. When World War II started he was deemed unfit for military service, and he instead worked in a factory making governors for propellers until the end of the war. He was an undergraduate at the University of Minnesota in 1946 when he was offered an opportunity to go to Alaska as a field assistant on a geologic reconnaissance expedition. Upon soliciting advice from the faculty, Sable was advised by the mineralogist, "You're crazy. You'll never graduate, Sable." Others were more supportive, and soon he was on his way north.

That year the party descended and mapped the terrain and geology of the Kurupa and Oolamnagavik rivers. For Sable it was the first of many extended forays in the Brooks Range and North Slope foothills. For more than a decade thereafter, Sable and other budding explorers would return to the region and spend entire summers exploring and mapping the geology. At age eighty-three, with no map in front of him, Sable can still name obscure peaks and creeks that they cataloged on those expeditions.

Sable and other scientists of his era were quick to acknowledge the exploration and scientific work that had preceded theirs. Most notable for Sable was a series of privately funded expeditions led by Ernest K. Leffingwell in 1907–1916. Working in remote areas of what is now the Arctic National Wildlife Refuge, Leffingwell explored the region, interacted with Eskimos on the coast, mapped the geology (including the coastline), and devised theories to explain the permafrost-dominated landscape. In 1948, Sable met the then-elderly Leffingwell by chance in George Gryc's office at USGS in Washington, D.C. Sable described Leffingwell as "really a pioneer, and his report is an excellent one—Professional Paper #109, I think it is—really got me going." Indeed it did.

In 1958 Sable repeated some of Leffingwell's 1907 glacier photographs and, with evidence from many surrounding moraines, showed that glaciers in the region were shrinking. It was one of the earliest uses of repeat

Ed Sable in his Alaska days (left), and in Denver, 2007 (right).

photography to assess landscape change. Sable put his early glacier findings in the context of the current changes. In 2007, he said:

> I think we are having a global, natural warming period. We've had several interglacial warming periods in the last couple-three centuries, so I feel that part of it is due to natural causes. At the time that I took these photos and all, I didn't have any real idea of global warming, except in the eastern Brooks Range. But, in looking at a lot of the glaciers, I had the strong feeling that they were all shrinking at the time. My own personal feeling is that man has accelerated this tremendously.

After getting a PhD for his Alaska work in 1962, Sable moved back to the Lower Forty-eight, where he worked as a geologist and raised a family on a large property in the foothills outside of Denver. He returned to Alaska briefly in 1984 to examine some old boreholes for coal potential, but from then on, most of Sable's ties to Alaska would consist of memories only. Perhaps each of us, upon aging, is prone to get a crack in the voice or a tear in the eye when reminiscing about old times, and Sable is no exception. Of all the terrain he had seen and rivers he had descended, he described his work in 1948 where the Jago, Okpilak, and Hula Hula rivers exit the mountains as being his favorite. "It was just a great year, because we were in beautifully scenic country. And of all those areas, the most scenic valley that I recall—and I'd like to see it again—was the Hula Hula River from the front. There were a great variety of rocks, from limestones to all sorts of things—basalts and so forth—and so I remember that very well. I'd really love to get back up there to take a look at the Okpilak and camp in the Hula Hula Valley, but…"

Sable (*right*) and his party in 1946. The notes for the photo read "L to R—Robert Thurrell Jr., geologist; Arthur Lachenbruch, camphand; Robert M. Chapman, geologist & chief of party; Delmar Isaacson, cook; Edward Sable, recorder. 9-3-46." That summer the party descended and mapped the Kurupa and Oolamnagavik rivers. ROBERT THURRELL, USGS

At Point Lay in 1947, during a visit by the USGS party that included Sable and Lachenbruch. Apparently a couple of Eskimos had been invited to see something or someone inside the tent. The party chief Ray Thompson snapped a picture of curious children. SABLE COLLECTION

Figure 27 (ABOVE) International Geophysical Year (IGY) camp on McCall Glacier, 1957. AUSTIN POST, USGS

Figure 28 (BELOW) Cross-section of McCall Glacier as the surface melts downward over time. The steep sides of the cross-section are the slopes of the moraine adjacent to the ice. The upper surface was inferred from 1956 USGS maps, hence the large vertical error. The other profiles were surveyed, and the error in these measurements is smaller than the symbols themselves. MATT NOLAN

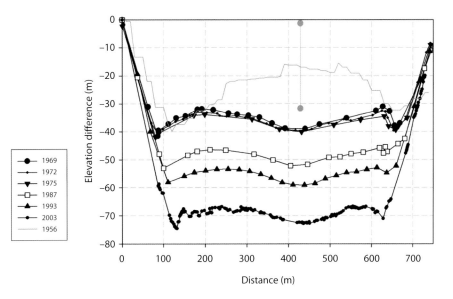

McCall Glacier had been the flagship U.S. glacier for the International Geophysical Year (IGY) in 1957–1958, and a Quonset hut had been erected high in the upper cirque (**Figure 27**) to serve as a center of operations. But the planned yearlong field effort ended tragically in November when leader Dick Hubley walked into the night and was later found dead. One of the tall peaks bears his name, but his name came up more often in connection with the "Hubley Fog" that rolls up the glacier like clockwork every afternoon. Hubley's group did manage to do some surveying of McCall Glacier prior to his death. Since that time, measurements of the glacier have become more refined, and there now exists a detailed record documenting the loss of ice since the 1950s (**Figure 28**).

Today the IGY camp from the upper cirque is melting out about a half mile from where it was in 1958. Anything deposited in the accumulation zone of a glacier, whether it be a lost screwdriver or an entire scientific camp, becomes buried by snow from ensuing years and entombed in ice until it melts out somewhere down-glacier in the ablation (melt) zone. Glaciers with a history of scientific research thus have a tendency to disgorge scientific instrumentation from earlier eras. Part of the responsibility of modern glaciologists is to retrieve these questionable treasures. For example, Nolan and I at one point in 2007 found ourselves headed back to our camp carrying a length of half-century-old rope and a couple cans of C-rations. Nolan later pronounced the rations to be less than tasty.

On a later outing to the lower cirque, en route to a 1973 black-and-white photo location near the terminus, I spotted an empty film box melting out of the ice. The box read "FUJI NEOPAN SS" and had a legible expiration date of January 1974 (**Figure 29**). This box had traveled through the glacier much like the IGY camp. It occurred to me that prints of photos from that roll might be those in my pocket.

Figure 29 Remains of film box found melting out of McCall Glacier. This box was dropped in the accumulation zone and then traveled down the glacier as the ice flowed. It eventually melted out at the surface, where it caught my eye.

Later that day I repeated some of Dennis Trabant's 1973 panoramic photos taken near the terminus (**Figure 30**). After reaching the photo location, I concentrated on the technical aspects of the photography before stopping momentarily to assess the change. It had only been thirty-five years since the initial photo, but the unnamed cirque glacier seen on the far side of McCall Glacier in the old photo had vanished—in a geologic instant!

I had already hiked up for a view of nearby Bur Cirque (not pictured) that afternoon, and I recalled how the sickly glacier looked like a pancake, plastered against but retracting from the underlying rock. Should the current climate trends continue, or even stabilize, Bur Cirque and hundreds of similar glaciers in the Brooks Range will disappear over the next several decades.

The same sort of shrinking can be seen in **Figure 30**, both in McCall Glacier in the foreground and in Hanging Glacier behind it and to the right. The thinning of McCall is especially striking; since the 1970s its thickness has decreased about 45 m on the lower glacier and about 10 m on the upper. Since the average thickness of McCall is 80 m, the documented thinning is substantial.

There are other indications that McCall is wasting away. On a glacier the snowline—the lower boundary of the snow still remaining from the previous winter—late in the summer roughly delineates the accumulation zone from the ablation zone for that particular year. So, if we are traversing a hypothetical valley glacier near the end of the summer, and if the glacier is in equilibrium with climate, then the first half of the uphill walk should be on exposed glacier ice (ablation zone), while the second half should be covered in snow (accumulation zone). Over the course of a year the lower half of the glacier is losing ice through summer melting, and the upper half is gaining ice through winter snowfall. But in equilibrium the glacier experiences no change in shape, because of the slow creep and flow of ice downhill.

Instead, on McCall Glacier in August 2007, we traversed the first 6.5 km on glacier ice before reaching the snow-covered ice only a kilometer from the top (**Figure 31**). That means, essentially, that mass is only being added to the glacier in that top kilometer; everywhere on the lower 6.5 km mass is being lost. Of course, a single year's observation might be rationalized as being anomalous, but glaciologists have been documenting the rising snowline on McCall for years. The snowline in some recent years has even been above the top of the glacier, indicating that no mass had been added that year; that is,

Figure 30 (OPPOSITE) Repeat photo pair from the eastern Brooks Range showing several glaciers shrinking. The old photo (top) is scanned and coarsely stitched from two 1973 prints (Dennis Trabant, USGS), and the new photo (bottom) is from August 11, 2007. Notice the drastic thinning of McCall Glacier that has occurred since the Little Ice Age trim lines were formed (top of the light-brown band). The cirque glacier on the left has disappeared, and the one on the right has shrunk. The old photo may have been taken earlier in the summer than the new one, but the snow does not obscure the extent of the ice mass.

Figure 31 (BELOW) A fifty-five-gallon drum on McCall Glacier, deformed from being buried in the accumulation zone a half-century earlier, traveling through the glacier, and reemerging in the ablation zone. Nearly all of the glacier in this view has exposed glacier ice at the surface, not last year's snow, indicating that only the very top of the glacier was accumulating mass this year. The IGY camp was between the two mountains pictured, near the top of the glacier.

Figure 32 Repeat photo pair from the eastern Brooks Range, showing unnamed retreating glaciers. The old photo (top), from the Col photo collection, is from August 20, 1950, and the new photos (bottom) are from August 6, 2007.

all of the previous winter's new snow had melted by the end of the summer. Meanwhile glacier ice was being lost through melt as usual.

By Nolan's estimate, even if the climate stabilizes immediately, McCall Glacier will likely lose the lower three-quarters of its current length before reaching an equilibrium with climate. It would then have a 1 km accumulation zone, consistent with last summer's climate, and about a 1 km ablation zone. Even under this optimistic and conservative climate scenario (stability), McCall would probably be one of only a dozen or so glaciers remaining in the Brooks Range, a number greatly reduced from the hundreds there presently.

Widespread Ice Loss

Although our group had trekked to the McCall Glacier from the coastal plain, rather than flying, we nevertheless had sporadic helicopter support once we reached the glacier. Near the end of our field season, Nolan and I boarded the helicopter to repeat some old photos and to see if the signs of shrinking on McCall were present in neighboring valleys as well. Under a lowering cloud ceiling and scattered showers, we flew past mountain peaks. It was immediately clear from the exposed glacier ice near the peaks that the snowline was above most peaks. Again, this means that most glaciers had no leftover snow from last year's winter, so that no mass was added to the glaciers this season. Furthermore, there were dozens of hanging and cirque glaciers of similar geometry and size to Bur Cirque. Given the current trends, we doubt that these glaciers will survive another thirty-five years.

I had long been seeking to repeat one particular fine-art-quality Col photo. On August 20, 1950, in a serendipitous photographic moment, an aerial photographer behind a military reconnaissance Fairchild camera had captured a perfect exposure of about ten hanging, cirque, and valley glaciers. I say "serendipitous" because the photos were being taken in rapid succession, with less than a second between exposures. Nonetheless, the photo is a scientific, compositional, and technical masterpiece. It has depth, detail, and dynamic range, and it demonstrates a variety of glacial features. Someone,

apparently as impressed as I was, has in the intervening years annotated various features on the photo.

From our helicopter we were getting only occasional glimpses of the mountain peaks, which were nearly lost in cloud. Unfortunately the peaks were exactly where we needed to be; the original photo had been taken from a fixed-wing airplane flying much higher than we were. As we swung into the narrow valley, I suggested to the pilot to go "up to the left, high in that cirque, next to the peak in the mist," but I knew that such a request was not likely to be granted. And I was right—my dream of repeating the photo slipped away. I realized, though, that I could perhaps use separate repeat photos—taken from the helicopter but from elevations lower than the original—to document the changes in the glaciers, and that with multiple photos I might even achieve a reasonably good repeat perspective on each glacier. I asked the pilot to swing around and fly down the valley, hoping that, as we flew by each glacier, I would for an instant be on a line between the glacier and the camera position of the old photo. The strategy worked well enough (**Figure 32**), and it saved me the embarrassment of repeating the exact old photo and not being able to equal the half-century-old technology! This is one of the most heavily glaciated parts of the Brooks Range, and the consistent radical shrinking in the repeated glacier insets leaves little doubt as to the trajectory of glaciers in the area (also **Figures 33 and 34**).

Future Monitoring of Brooks Range Glaciers

The most important number associated with a glacier is its volume. In order to measure future changes in glacier volume, digital elevation models (DEM: precise surface elevation map) for most of the 842 glaciers in the Brooks Range have now been acquired. When DEMs are reacquired in the future, comparison of current and future DEMs will allow ice volume gains or losses to be calculated. The precise, yet widespread, monitoring of Brooks Range glacier volume available today is an achievement that would have been almost unimaginable in the days of Leffingwell and Sable.

1957. Looking south Small glacier on south tributary of McCall Creek.

Figure 33 Repeat photo pair from the McCall Creek valley. The old photo (top) is from 1957 (Ed Sable), and the new photo (bottom) is from July 30, 2007. Handwriting from Sable's notes reads "1957. Looking south Small glacier on south tributary of McCall Creek." None of the previous winter's snow remains on the glacier in the new photograph, meaning that no mass was added to the glacier that year. Should warming trends continue, this glacier, like many others its size in the Brooks Range, will disappear before the next half-century interval elapses.

 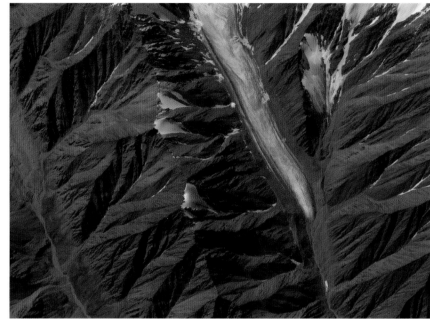

Figure 34 Repeat photo pair from the eastern Brooks Range, showing unnamed retreating glaciers. The old photo (left) is from July 16, 1947, and the new photo (right) is from July 24, 2006. All of the bright areas in both photos are glacier ice. SABLE COLLECTION; GOOGLE EARTH

Permafrost

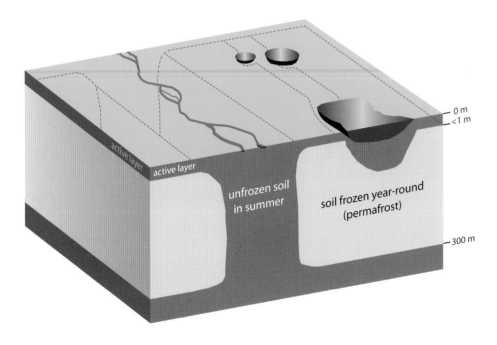

Figure 35 Idealized permafrost distribution diagram, showing the block of frozen soil, continuous except under some rivers and lakes.

GENERALLY SPEAKING, GROUND STILL frozen at the end of two summers is considered permafrost. More precisely, permafrost is ground that has been at or below 0°C (32°F) for at least two years. Most of the North Slope of Alaska is underlain by permafrost that begins within one meter of the surface and extends several hundred meters downward.[9] The layer of soil above the permafrost that freezes and thaws seasonally is called the active layer (**Figure 35**).

If the temperatures at the top and bottom of the permafrost were to remain constant over a long period, then the permafrost would ideally be expected to have a constant temperature gradient. That is, the change (increase, actually) in temperature for a given change in depth should be the same everywhere; if temperature is plotted as a function of depth, the result should be a straight line (**Figure 36**).

In the Arctic, the temperature is indeed thought to be constant at the bottom of the permafrost. The surface temperatures, meanwhile, fluctuate on diurnal, seasonal, decadal, and longer time scales. Seasonal fluctuations are detectable up to 18 m into the permafrost; decadal and longer fluctuations can penetrate deeper. Thus the permafrost retains a memory of climate change in the portion of its vertical temperature profile below 18 m. A linear (i.e., straight) profile there indicates climate stability, probably over a period of a century or longer. By identifying nonlinear (curved) temperature profiles deeper than 18 m into the frozen ground, we can infer decadal and longer warmth anomalies at the surface. A larger warming produces larger departures from the linear profile.

The Imprint of Warming on Permafrost

Art Lachenbruch was a pioneer in the science of permafrost. In 1986 Lachenbruch published findings that showed the signature of warming arctic air temperatures imprinted on the temperature profiles in deep permafrost boreholes scattered across the North Slope (**Figure 37**). The boreholes had been drilled for oil exploration purposes and later capped. Temperatures going down into the boreholes were measured immediately after drilling, and then multiple times thereafter. The subsequent cooling, in Lachenbruch's words, "follows the simple heat conduction model for a line source (the borehole) of finite duration (the momentary drilling) in an infinite homogenous medium (the soil)." The declining rate at which the heat from drilling dissipated into the surrounding frozen soil allowed Lachenbruch to infer the thermal conductivity of the material. He was then able to calculate the desired (equilibrium) temperature profiles that had existed before the heating and disturbing of the permafrost with drilling.

Much earlier—in 1959—Lachenbruch had published his PhD findings, including evidence from a borehole near Barrow that contained a deflected temperature profile indicative of recent warming.[10] The manuscript used early permafrost borehole evidence to infer the warming trend from the 1930s and 1940s. In 1959, he wrote:

> This work forms an adequate basis for the study of secular changes in ground temperature that are measurable during

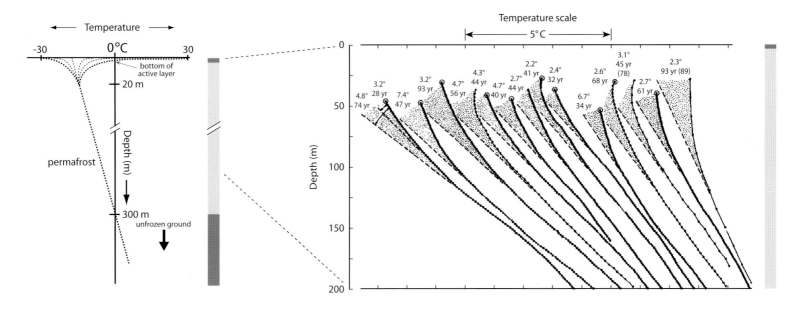

Figure 36 Temperature of soil substrate, as it changes with depth below surface, assuming only seasonal temperature fluctuations are present. The trumpet-shaped dashed curves at the top represent the temperature profiles at different times during the year.

Figure 37 Temperature as a function of depth in fourteen boreholes from northern Alaska, 1986.[11] The horizontal axis is a relative temperature scale, and the curves are plotted next to each other, so that the similarity between the borehole temperature measurements can be seen. The solid curves represent the actual measured borehole temperatures, and the dotted lines represent what the temperature should be, were the mean annual surface temperature not increasing. The positive deflection of the solid curves from the dotted lines reveals that surface temperatures had been increasing over the previous decade and century. Numbers above each curve denote the duration and magnitude of the inferred temperature increase. From Lachenbruch, A. H., and B. V. Marshall (1986). Changing climate: Geothermal evidence from permafrost in the Alaskan Arctic. *Science* 234–4777:689–696. Adapted and reprinted with permission from AAAS.

Figure 38 Same as Figure 11 but with modern close-up photos of two foreground rocks seen in the old photo. Both rocks have migrated downhill. The left rock has retained its initial orientation. The rock on the right has been forced upright by the large willow shrub growing underneath it. Also, the steep slope to the right of the figure in the old photo has collapsed as the underlying permafrost has thawed. Three large boulders in the old photo are also marked in the new photo. Boulder #1 barely moved during the interval between photos, but changed its orientation somewhat. Boulder #2, which had just begun to slide in the old photo (note dark scar above boulder), slid down the collapsing slope, while maintaining its orientation. Boulder #3 rolled down and changed its orientation.

the period of observation. By correcting the observed temperatures for the effects of drilling, it is possible to detect residual secular changes in natural earth temperatures as small as 0.01°C per year long before thermal equilibrium is restored. Changes of this order have been observed in the upper portion of the depth interval studied. Preliminary results suggest that such changes correspond to an increase in temperature at the ground surface on the order of 3°C during the past 50 to 75 years.

Wrinkles in the Landscape

Thawing of ice-rich permafrost can create voids in the ground that then lead to subsidence. The resulting depressions in the ground surface can range in size from a few centimeters up to 20 m wide and 10 m deep. Irregular topography that arises from thawing of ice-rich permafrost is known as thermokarst.

It is thermokarst that I think we are seeing in the photo pair of **Figure 38**. The photo pair is striking, so let me elaborate on the circumstances. The old photo came from Ed Sable's field photo collection and contained the note, "Near top of moraine W side of Jago River." Well, the Jago River is about 150 km long, and most of its length has been glaciated at one time or another, leaving no shortage of moraines, so there was a huge range of possible photo locations. But, although the old photo is foggy, I saw that it did have detailed foreground and identifiable background—both necessary ingredients for a successful repeat. The barely visible distant mountains, along with the moraine hill and the lake in the middle ground, allowed me to guess the general location of the photograph.

When I arrived at the location, the foreground appeared so radically different from the foreground in the old photo that I wondered whether this was really the correct location or whether I was still a little bit off. And where were those two distinctive rocks that appeared in the foreground of the old photo? Surely they couldn't have just disappeared. I probed around among the head-high willow shrubs, and there were the two rocks (**Figure 38**). But when I positioned the camera so that the rock locations in my viewfinder would mimic those in the old photo, I could no longer see the lake that appears in the old photo (to the right of the hill in the center). The rocks had apparently crept down the side of the gully. Their initial orientations had

barely changed, suggesting that they did not roll, but rather slid down the slope, as might be expected in a permafrost landscape.

Despite my best efforts, I never got the repeat perspective quite right. I had plenty of time to adjust my position and attempt to recreate the exact view seen in Sable's old photo (except for the two rocks, of course), but I could never attain the proper triangulation between the near and middle ground. In the end there was little doubt about where my camera needed to be, but the topography of the gully had changed, putting the needed camera position out of reach, just above the gully. The gully had apparently widened.

The photo pair (**Figure 38**) shows how rapidly the gully has been collapsing. In the old photo, the right side of the gully is a steep slope with blocks of tundra and several large boulders on it. In the new photo, the steep slope has thawed and the entire surface has subsided. Three large boulders in the old photo can be identified in the new photo.

It is a remarkable photo pair, with its collapsing terrain and exploding vegetation. What are we to make of it?

The foreground terrain in the photos is glacial moraine. But it is not recent and in fact it is thought to be at least nine thousand years old.[12] So it is not as if Sable was standing on ground recently vacated by some nearby glacier—there are none—and that shrubs have naturally moved in to take advantage of the newly available space. Instead, one cannot help wondering whether the area might have been shrub-free for several thousand years before Sable's arrival. In that case, the changes documented in the photo pair are much more dramatic, perhaps even alarming, since they represent such a radical departure from what had (presumably) preceded. But of course I cannot prove that things were not changing before Sable; I cannot prove, for example, that shrubs had not come and gone multiple times and that Sable had not coincidentally arrived when shrubs were at their low point in a cycle.

Nor can I prove that past episodes of thermokarst development had not occurred, but in this case presumably those episodes would have left behind evidence in the form of irregular topography, which is uncommon on this moraine. It seems that the gully is one of the first major incisions in the old moraine since the ice retreated thousands of years earlier. If only Leffingwell, fifty years before Sable, had made it to the same spot with his camera! It is hard to compute a trajectory when you have only two data points to work with.

Art Lachenbruch

From interview with the author on December 14, 2007:

Art Lachenbruch accompanied Ed Sable on several early adventures on the North Slope and Brooks Range, as part of the geologic reconnaissance described in the text. A self-described weak student in high school, Lachenbruch was recruited to analyze military trimetrogon reconnaissance photography. He told the recruiter, a veteran of many trips to Alaska, that it sounded interesting, but that what he really wanted to do was get up to Alaska. Soon thereafter, at age seventeen, Lachenbruch was sent to Alaska as a backpacker and cook for a summer field party. After returning, he enlisted and fought overseas in World War II. While abroad, he ordered reprints of standard college texts through the Armed Forces Institute and passed tests granting him college credit upon his return home. However, most colleges would still not accept him. But some of his field buddies from Alaska had connections at Johns Hopkins, and so it was that Lachenbruch entered college. At least, that is Lachenbruch's self-deprecating version of the story.

Outside of the nearly inaccessible Soviet literature, little was known about permafrost at the time, but Lachenbruch would change that. During his undergraduate years at Johns Hopkins and his graduate years at Harvard, Lachenbruch spent six successive summers in northern Alaska. The principal goals of the field efforts were to map the bedrock geology of the region, but he was often granted time to explore his interest in permafrost features. He soon encountered Leffingwell's epic USGS Professional Paper 109, which, as Lachenbruch said, "enhanced, if not inflamed, my interest," and this led to a PhD on heat transfer in frozen ground.

During his tenure in Alaska, Lachenbruch became perhaps the preeminent permafrost scientist in the country and was consequently called to

Lachenbruch packing caribou, May 22, 1946. USGS

Art Lachenbruch in northern Alaska, late 1940s, and today.
RAY THOMPSON, COURTESY OF ED SABLE; COURTESY OF ART LACHENBRUCH.

evaluate the potential impact of proposed engineering projects on frozen ground. At the same time that he was publishing his dissertation research, he was on the environmental committee for Project Chariot. Project Chariot was a plan to explode buried atomic bombs to excavate a harbor in northwestern Alaska for shipping. Although Project Chariot was eventually blocked, for many good reasons, the product of the environmental assessment, titled "Environment of the Cape Thompson Region, Alaska," laid the foundation for the National Environmental Policy Act (NEPA), and was a predecessor to the modern environmental impact statement (EIS).

Later in his career, Lachenbruch caught wind of a proposed oil pipeline, and he told the director of USGS that if U.S. Secretary of the Interior Walter Hickel should ask the director for an opinion on it, not to tell Hickel anything until they had had a chance to make some calculations. In fact, the director had already been contacted, and the clock was ticking. Lachenbruch was granted one week to determine the effect of burying an oil pipeline across a permafrost region. The thermal requirements that Lachenbruch deemed necessary for a lasting pipeline raised the price tag from one billion to nine billion dollars. Suddenly economics and permafrost science were inextricably tied, and it was at that moment that frozen ground captured the attention of a broader audience. Unlike Project Chariot, the Trans-Alaska Pipeline System (TAPS) did come to fruition. The construction requirements set forth by Lachenbruch have allowed the frozen ground and the pipeline—sometimes buried and sometimes elevated—to coexist for three decades. Oil in the pipeline was the ultimate finish line to the work that Gryc had started years earlier on a sandbar of Prince Creek.

Like Sable, Lachenbruch fondly recalled his early days in Alaska. About the region, he said it was "fantastic country...it's all fantastic." The science was really what got him going, though. "My main effort was trying to understand the geomorphic features, particularly mechanical processes, that take place in a permafrost terrain, that are unique to permafrost, and how they affect the workings of the countryside....For me, the most remarkable thing that seemed to transcend the whole countryside in that area was the evolution of ice wedge polygons and their enormous geomorphic effects."

(ABOVE) Left to right: Mickey Toorak, Charlie Tuckfield, and Leo Atungaruk greet Sable, Lachenbruch, and party at the mouth of the Utukok, for escort to Point Lay, 1947. SABLE COLLECTION

(BELOW) Sable and Lachenbruch (barely visible behind Sable) loading their canvas boat, with camp hand Charles Marrow and cook Ernest Wadsworth looking on. The party is descending the Utukok River during spring breakup, 1947. SABLE COLLECTION

The collapsing terrain here (**Figure 38**) is likely a consequence of the thawing of the permafrost below. If I had to guess, I would say that it is a sign of some climatic change: warming, deeper snow, or more severe rains. For example, snow insulates the ground during the winter and slows the loss of heat from the ground to the air. Put more intuitively: the snow slows the penetration of cold from the air to the ground. Some permafrost scientists believe that increases in snow depth thus play a large role—perhaps as large as increases in air temperature—in raising the temperature of the permafrost and causing thermokarst.[13,14] In fact, once thermokarst begins to develop, the depression traps blowing snow. The deeper snow then warms and thaws additional permafrost, and the depression continues to deepen.

Whatever is causing the changes in the photo pair, similar changes seem to have happened nearby (**Figure 39**), and elsewhere on the North Slope.[15] The ancient moraine is covered with small troughs and microrelief, like much of the North Slope. This area can be viewed in Google Earth (spherical panoramas under "Gallery > 360 Cities").

The photo pair (**Figure 38**) also suggests that there is a connection between surface depressions and shrubs (**Figure 40**). Indeed, thermokarst topography and microrelief are conducive to shrub expansion. People who have spent time in the Arctic have observed that shrubs often grow where there are depressions. "Where there are wrinkles, there are shrubs."

Figure 39 Recently formed depression in the surface. This is near the widening gully in Figure 38.

↓ small shrubs   – – – – – depressions

Figure 40 Ancient (9,000+ ya) glacial moraine. There is a strong correlation between the depressions (dotted lines), which are thought to be deepening, and the shrubs (yellow arrows). From studying patterns present in many repeat photo pairs from other localities, we think we can reasonably guess what a landscape such as the one here would have looked like in 1950, perhaps even in 1850. The depressions would have been less pronounced, there would have been fewer shrubs, and more lichens would have been growing where grasses now thrive.

Implications

I HAVE PRESENTED SOME rather solid evidence for change in the landscape of northern Alaska. The changes are consistent with a warming climate, but the argument for warming is not as solid as the argument for the changes themselves. In any case, there is another reason why some scientists care—and even worry—about the changes.

In the past, complicated feedback mechanisms apparently caused most of the large changes in temperature between glacial and interglacial periods. Many scientists recognize that these feedbacks are lurking within the arctic system, ready to transform today's small and seemingly innocuous changes into unpredictable and perhaps disastrous changes in the future.

Sea ice, mentioned in the introduction, provides a relatively simple example of (positive) feedback. Bright reflective sea ice melts and is replaced by the dark absorbent ocean. The dark ocean retains more of the sun's radiation than the bright sea ice, so the ocean warms. Warming leads to the melting of additional sea ice, which in turn leads to more warming. Thus the result of warming is . . . even more warming. Things could run amok.

On land, the feedback mechanisms are not always intuitive. One might guess, for example, that shrubs would inhibit warming, since they remove CO_2 (carbon dioxide) from the atmosphere (for photosynthesis) and thus diminish the greenhouse (warming) effect of the atmosphere. In fact, it seems not to work that way. Let's agree, for the sake of the argument, that warming leads to shrub expansion. So warming means more shrubs. Where there are shrubs, they trap snow and thus warm the ground. The warmer soil increases winter decomposition of carbon in the soil.[16] Where there are shrubs, there is also less soil carbon.[17] Both correlations suggest that shrub expansion will release carbon from the soil into the atmosphere—more than enough, in fact, to offset the CO_2 removed from the atmosphere in photosynthesis.[18] More atmospheric CO_2 then leads to even faster warming. Thus, instead of inhibiting warming, shrubs might actually enhance warming, and they might at some point begin to do so disastrously.

One can argue about the details of the preceding reasoning, but the point is that, with climate-landscape interactions, we are dealing with a very complex system, one that may contain dangerous instabilities that we humans do not want to trigger. Decision makers need to err on the side of caution.

But my intent in writing this book was not so much to try to predict the consequences of the change that we are seeing in the arctic landscape, but rather to present some evidence for change and then to let the scientific community as a whole grapple with the implications.

The terrestrial Arctic is a complex system that is yet to be fully understood. Until a better understanding emerges, the specific climatic causes of the changes presented here will likely be debated. The changes in these repeated photos, however, are unequivocal. A visitor to the Arctic may be struck by the seeming timelessness and constancy of the place, but that impression is misleading. The arctic landscape is changing.

Figure 41 Tussocks regrowing two years after a rare large (1,000 km² or 250,000 acre) tundra fire near the Anaktuvuk River, North Slope. It is unclear if the large fire is a harbinger of more tundra fires in the future. The introduction of a fire regime to the tundra would alter the landscape dramatically.

Locations of vegetation photo-pairs (WGS 84 datum)

		Latitude, N	Longitude, W
	Figure 11	69°27.104′	143°44.628′
	Figure 14	68°15.455′	159°54.683′
	Figure 15	68°52.181′	154°10.295′
	Figure 16	68°47.015′	152°01.792′
	Figure 17	67°57.272′	161°39.300′

		Latitude, N	Longitude, W
	Figure 18	67°57.510′	161°37.942′
	Figure 19	67°57.220′	161°40.263′
	Figure 20	67°57.727′	161°49.758′
	Figure 23	69°26.324′	143°46.354′
	Figure 24	69°05.313′	148°44.422′

Locations of glacier photo-pairs (WGS 84 datum)

		Latitude, N	Longitude, W
	Figure 26	69°10.723′	144°10.186′
	Figure 30	69°19.565′	143°51.865′
	Figure 32	69°10.206′	144°06.297′
	Figure 33	69°20.185′	143°45.652′
	Figure 34	69°10.800′	143°57.297′

References

1. NASA/Goddard Space Flight Center Scientific Visualization Studio.

2. Joly, K., R. R. Jandt, C. R. Meyers, M. J. Cole (2007). Changes in vegetative cover on Western Arctic Herd winter range from 1981 to 2005: Potential effects of grazing and climate change. *Rangifer* special issue 17:199–207.

3. Chapin III, F. S., G. R. Shaver, and A. E. Giblin (1995). Responses of arctic tundra to experimental and observed changes in climate. *Ecology* 76–3:694–711.

4. Goetz, S. J., M. C. Mack, K. R. Gurney, J. T. Randerson, R. A. Houghton (2007). Ecosystem responses to recent climate change and fire disturbance at northern high latitudes: Observations and model results contrasting northern Eurasia and North America. *Environmental Research Letters* 2:4531.

5. Briner, J. P., D. S. Kaufman, W. F. Manley, R. C. Finkel, and M. W. Caffee (2005). Cosmogenic exposure dating of late Pleistocene moraine stabilization in Alaska. *Geologic Society of America Bulletin* 117–7/8:1108–1120.

6. Leffingwell, E., (1919) The Canning River region, northern Alaska. USGS Professional Paper 109, 251 p.

7. Sable, E. G. (1961) Recent recession and thinning of Okpilak Glacier, northeastern Alaska. *Arctic* 14–3: 176-187

8. Arendt, A., K Echelmeyer, W. Harrison, C. Lingle, and V. Valentine (2002). Rapid wastage of Alaska glaciers and their contribution to rising sea level. *Science* 297:382–386.

9. Osterkamp, T. E., J. K. Petersen, and T. S. Collett (1985). Permafrost thickness in the Oliktok Point, Prudhoe Bay and Mikkelsen Bay areas of Alaska. *Cold Regions Science and Technology* 11:99–105.

10. Lachenbruch, A. H. (1959) Dissipation of the temperature effect of drilling a well in Arctic Alaska. Geological Survey Bulletin 1083-C, 109 p.

11. Lachenbruch, A. H., and B. V. Marshall (1986). Changing climate: Geothermal evidence from permafrost in the Alaskan Arctic. *Science* 234(4777):689–696.

12. Briner et al., op. cit.

13. Osterkamp, T. E., and V. E. Romanovsky (1999). Evidence of warming and thawing of discontinuous permafrost in Alaska. *Permafrost and Periglacial Processes* 10:17–37.

14. Osterkamp, T. E. (2007). Causes of warming and thawing permafrost in Alaska. *EOS* 88–48:522–523.

15. Jorgenson, M. T., Y. L. Shur, and E. R. Pullman (2006). Abrupt increase in permafrost degradation in Arctic Alaska. *Geophysical Research Letters* 33 – L02503

16. Fahnestock, J. T., M. H. Jones, P. D. Brooks, D. A. Walker, J. M. Welker (1998). Winter and early spring CO_2 efflux from tundra communities of northern Alaska. *Journal of Geophysical Research* 103:29023.

17. Mack, M. C., E. A. G. Schuur, M. S. Bret-Harte, G. R. Shaver, and F. S. Chapin III (2004). Ecosystem carbon storage in arctic tundra reduced by long-term nutrient fertilization. *Nature* 431:440–443.

18. Ibid.

Index